WHEN I GO TO THE ZOO, WHAT DO I SEE?

Miranda Kelly

CRABTREE
PUBLISHING COMPANY
WWW.CRABTREEBOOKS.COM

When I go to the **zoo**, what do I see?

ZOO
ENTRANCE
BUY TICKETS
NOW OPEN
PURCHASE TICKETS HERE
BUY TICKETS
6 OPEN
8 OPEN
Your Membership
PAYS FOR ITSELF
IN AS LITTLE AS

I see a **giraffe**, tall as a tree.

TREE
GIRAFFE

I see funny **monkeys**,
one, two, three.

1
2
3

I see an **elephant** with her baby.

ELEPHANT
BABY ELEPHANT

I see a **polar bear**,
white as can be.

I see a bold tiger staring at me.

When I go to the zoo, what do I see?

I see a **hippo** and fish in a pond of green.

I see goats and rabbits we get to feed.

I see wonderful animals. Don't you agree?

Glossary

elephant (EL-uh-fuhnt): An elephant is a large animal with tough, wrinkled skin and a trunk.

giraffe (juh-RAF): A giraffe is an animal with a very long neck. It is the tallest animal in the world.

hippo or hippopotamus (hip-uh-POT-uh-muhss): A hippopotamus is an African animal with short legs and thick skin. Hippos live in or near water.

monkeys (MUHNG-keez): Monkeys are animals that are like small apes. Monkeys usually have a tail.

polar bear (POH-lur BAIR): A polar bear is an animal that comes from the cold Arctic.

zoo (ZOO): A zoo is a place where animals are kept for people to see.

Index

School-to-Home Support for Caregivers and Teachers

Crabtree Seedlings books help children grow by letting them practice reading. Here are a few guiding questions to help the reader with building his or her comprehension skills. Possible answers are included.

Before Reading

- What do I think this book is about? I think this book is about visiting a zoo.
- What do I want to learn about this topic? I want to learn about the animals found at a zoo.

During Reading

- I wonder why... I wonder why the hippo is under the water.
- What have I learned so far? I have learned that giraffes, monkeys, elephants, polar bears, tigers, hippos, fish, goats, and rabbits are found at a zoo.

After Reading

- What details did I learn about this topic? I learned that there are many different animals at zoos. People can learn about the animals from zoo workers. People can even feed some animals.
- Read the book again and look for the vocabulary words. I see the word ***monkeys*** on page 6 and the word ***hippo*** on page 16. The other vocabulary words are found on pages 22 and 23.

Library and Archives Canada Cataloging-in-Publication Data

Title: When I go to the zoo, what do I see? / by Miranda Kelly.
Names: Kelly, Miranda, 1990- author.
Description: Series statement: In my community | "A Crabtree seedlings book". | Includes index.
Identifiers: Canadiana 20200388118 | ISBN 9781427129628 (hardcover) | ISBN 9781427129727 (softcover)
Subjects: LCSH: Zoo animals—Juvenile literature. | LCSH: Zoos—Juvenile literature.
Classification: LCC QL77.5 .K45 2021 | DDC j590.73—dc23

Library of Congress Cataloging-in-Publication Data

Names: Kelly, Miranda, 1990- author.
Title: When I go to the zoo, what do I see? / by Miranda Kelly.
Description: New York : Crabtree Publishing, 2021. | Series: In my community : a Crabtree seedlings book | Includes index.
Identifiers: LCCN 2020050759 | ISBN 9781427129628 (hardcover) | ISBN 9781427129727 (paperback)
Subjects: LCSH: Zoo animals--Juvenile literature.
Classification: LCC QL77.5 .K45 2021 | DDC 590.73--dc23
LC record available at https://lccn.loc.gov/2020050759

Crabtree Publishing Company
www.crabtreebooks.com **1-800-387-7650**
e-book ISBN 978-1-947632-59-2
Print book version produced jointly with Crabtree Publishing Company NY, USA

Written by Miranda Kelly
Production coordinator and Prepress technician: Ken Wright
Print coordinator: Katherine Berti

Printed in the U.S.A ./012021/CG20201112

Photo credits: Cover background, text box art, and tickets page 2 © shutterstock.com/mollicart; page 3 © shutterstock.com/Trong Nguyen; page 3 © istock.com/stevezmina1; page 4 © Mrs_ya; page 5 © shutterstock.com/Susan Macdonald; page 7 © shutterstock.com/yod67; page 9 © shutterstock.com/pitchakorn yodmun; page 11 © shutterstock.com/Marina_Saw_it; page 13 © shutterstock.com/iyd39; page 15 © istock.com/image source; page 17 ©istock.com/tatisol; page 19 ©istock.com/Andrey Donnikov.

Published in Canada
Crabtree Publishing
616 Welland Ave.
St. Catharines, ON
L2M 5V6

Published in the United States
Crabtree Publishing
347 Fifth Ave
Suite 1402-145
New York, NY 10016

Published in the United Kingdom
Crabtree Publishing
Maritime House
Basin Road North, Hove
BN41 1WR

Published in Australia
Crabtree Publishing
Unit 3 – 5
Currumbin Court
Capalaba QLD 4157